Name: _____

MAGIC NUMBERS
STUDENT BOOK 1

A. A. Sarkiss

Printed by CreateSpace, An Amazon.com Company

Available from Amazon.com, CreateSpace.com and other retail outlets

Download Magic Numbers: Parents' and Teacher's Guide 1
http://magicnumbers.yolasite.com

1 2 3 4 5

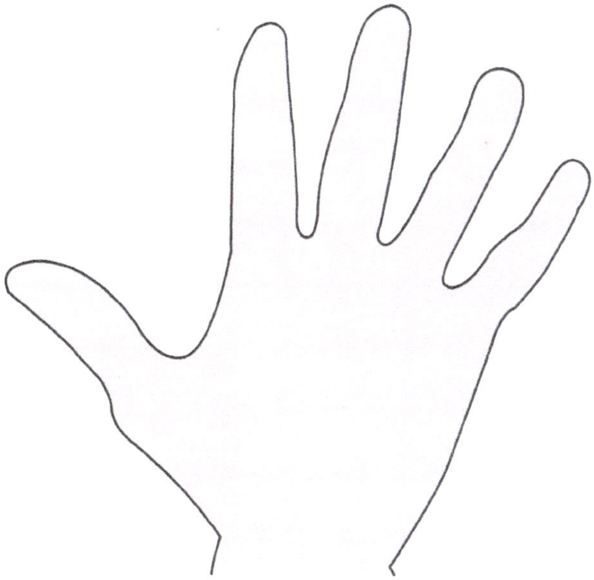

★ Count

1

2

3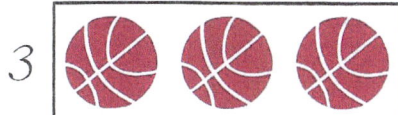

4

5

★ Read and circle

1

2

3

4

5

1

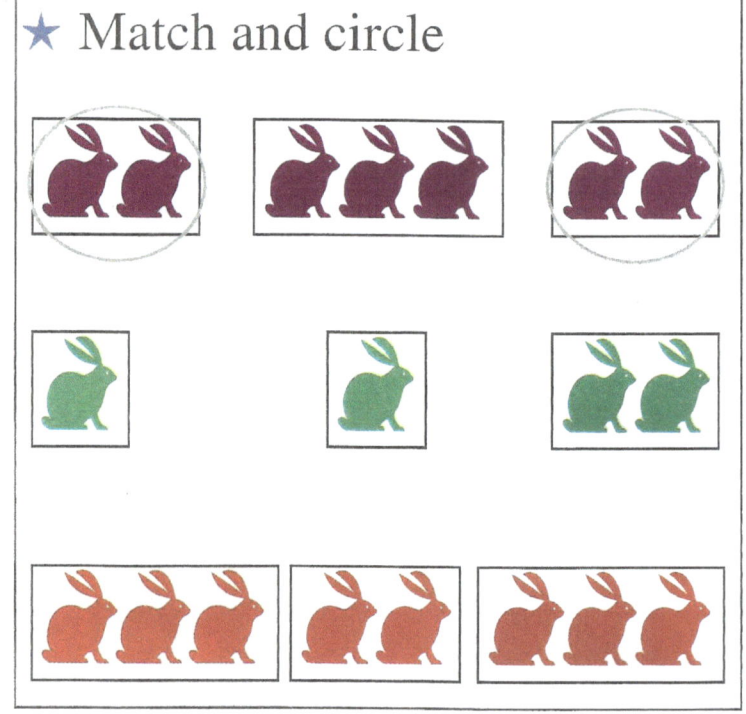

★ Read and write

★ Read and circle   ★ Match and circle

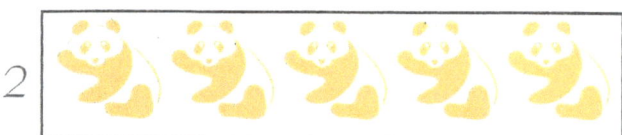

2

★ Read and write

2

★ Read and circle

1

3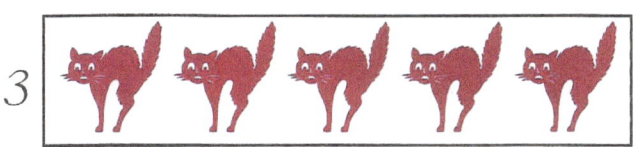

2

★ Match and circle

3

★ Read and write

3

★ Circle the correct number

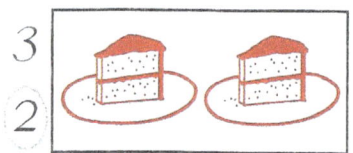

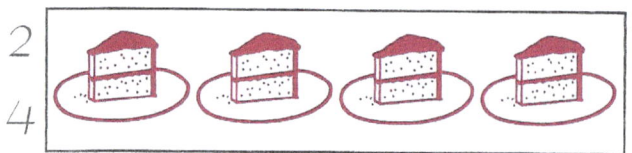

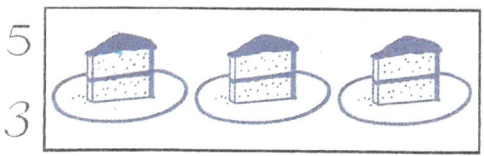

★ Count and colour

Count and Colour

blue	green	red	yellow	orange
*	**	***	****	*****

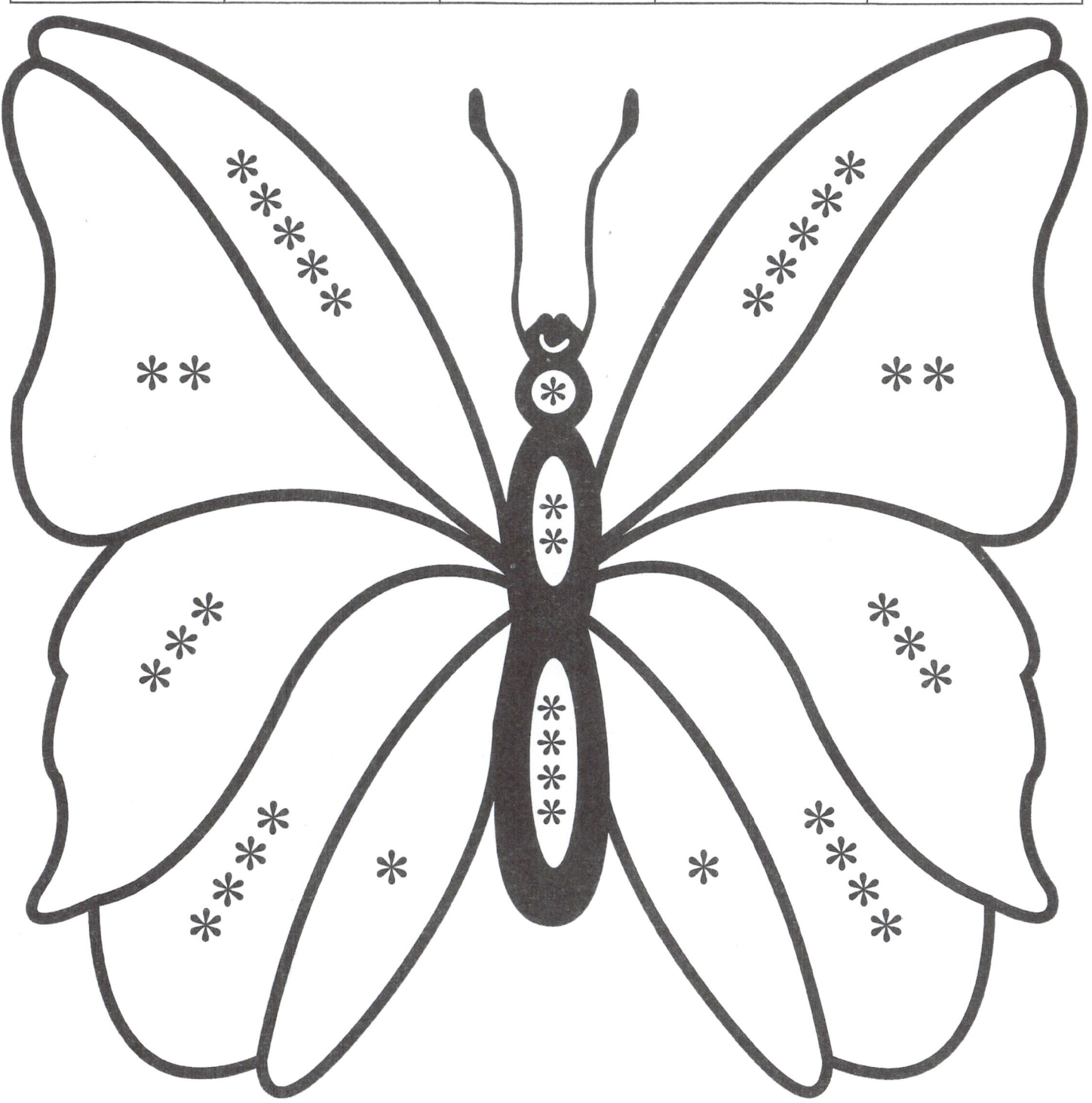

Date: ....

4

★ Read and write

★ Circle the correct number

4
3

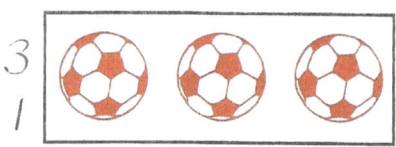

3
1

4
5

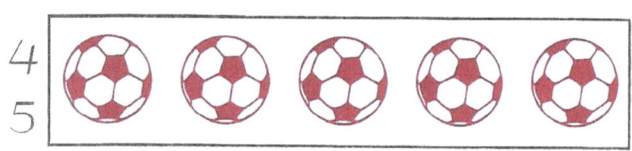

★ Count and colour

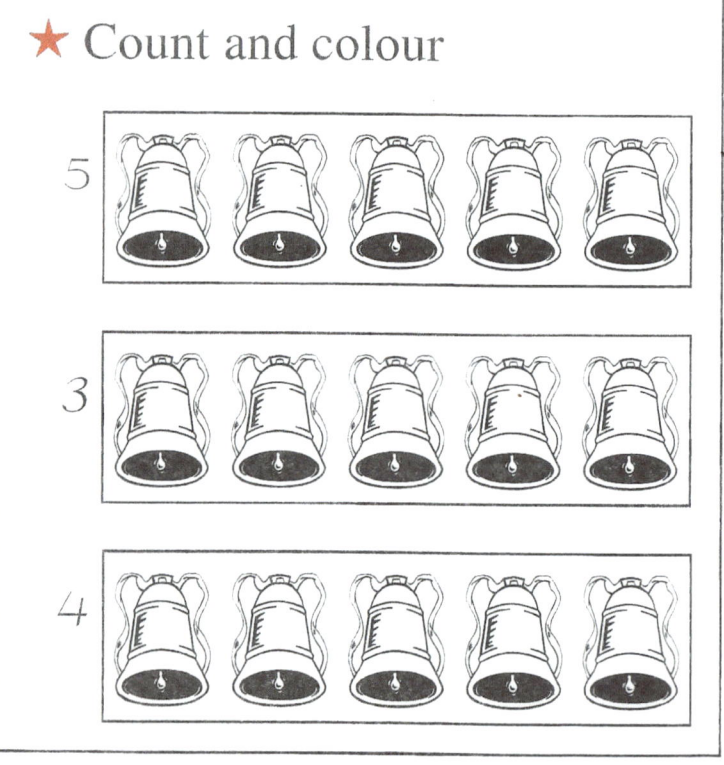

5

3

4

Count, colour the boxes and write

5

⭐ Read and write

⭐ Read and circle   ⭐ Match and circle

3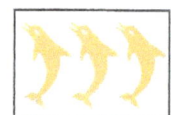

5

4

Find, count, colour the boxes and write

Revision

★ Read and circle

★ Match and circle

★ Circle the correct number

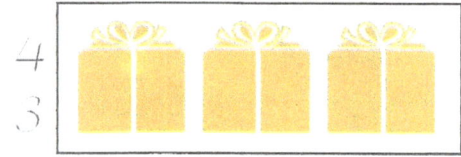

★ Count and colour

1 2 3 4 5 6 7 8 9 10

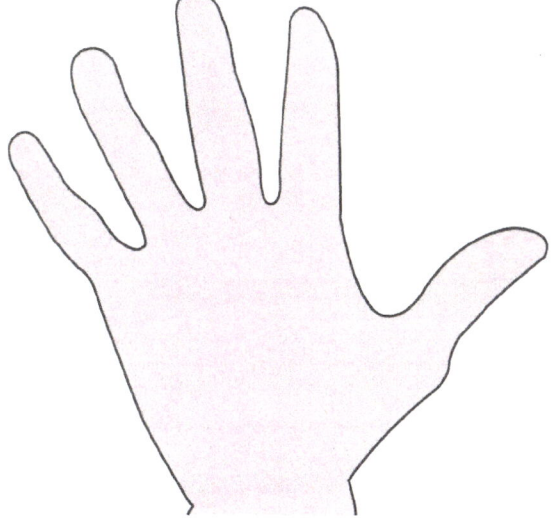 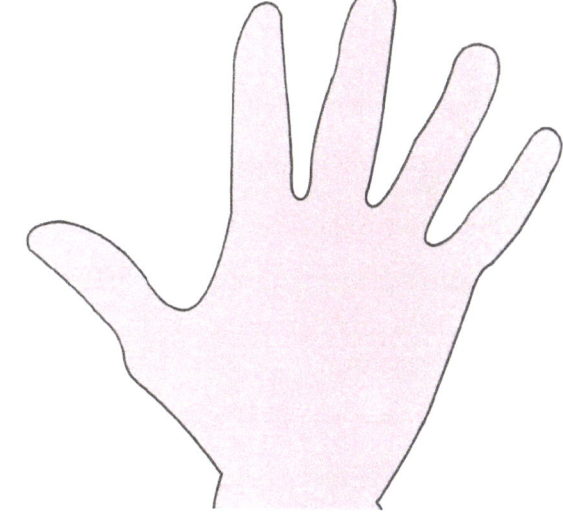

★ Count

6

7

8

9

10

★ Read and circle

6

7

8

9

10

6

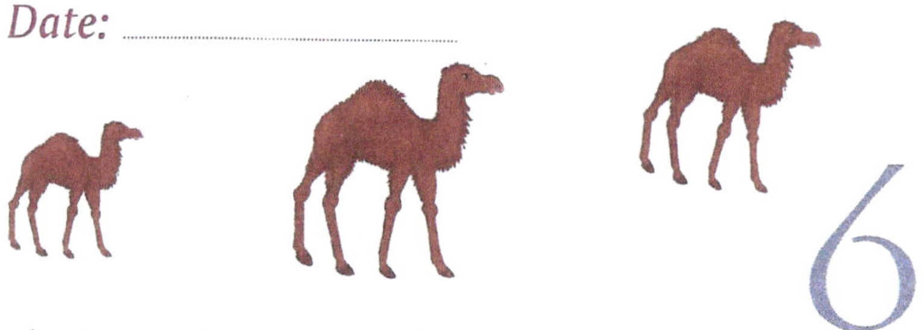

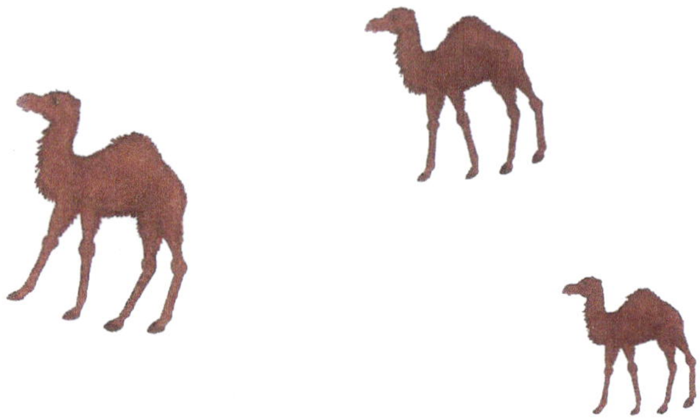

★ Read and write

6

★ Read and circle

7

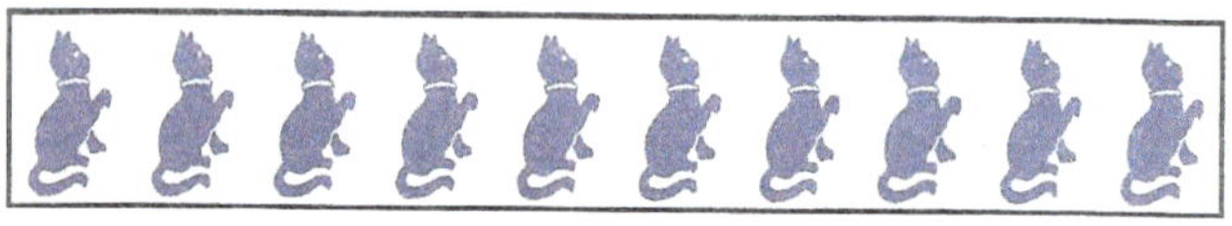

8

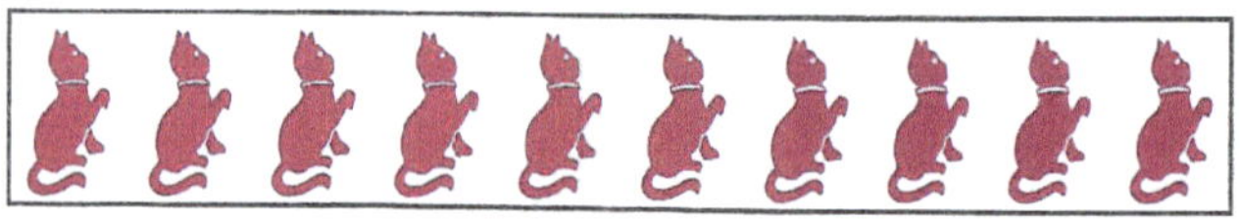

5

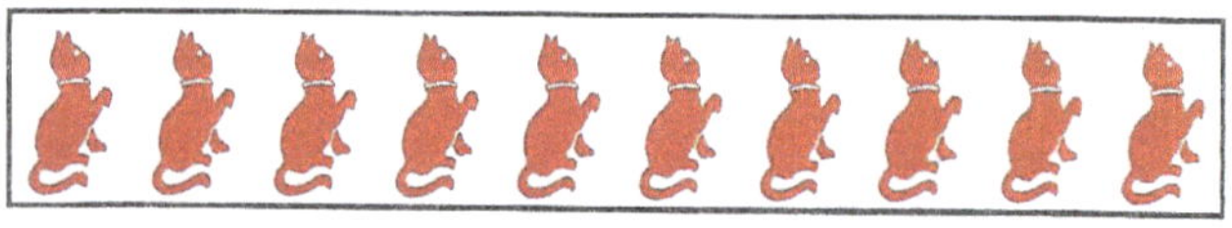

6

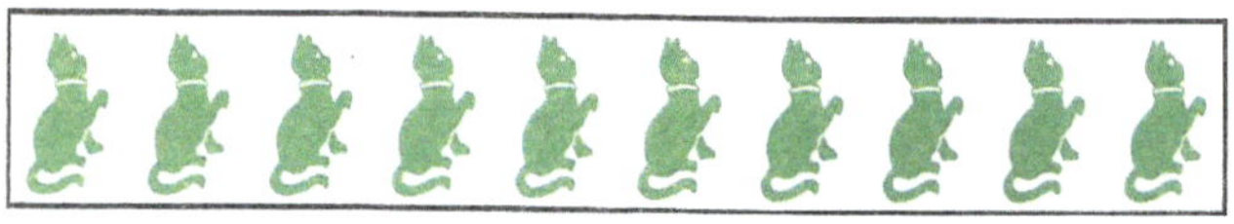

★ Read and write

Date: ....

Find, count, colour the boxes and write

★ Read and write

★ Match and circle

★ Write

★ Read and write

8
8

★ Circle the correct number

6
5

8
5

7
9

7
8

★ Write

Count and match

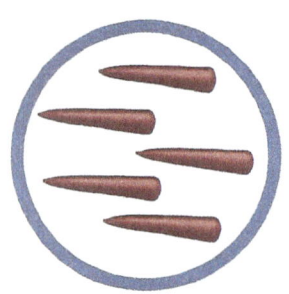

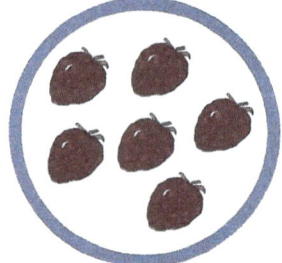

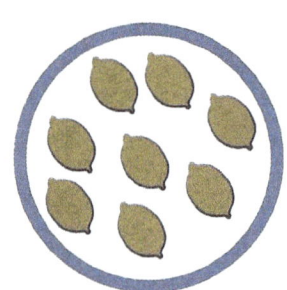

9

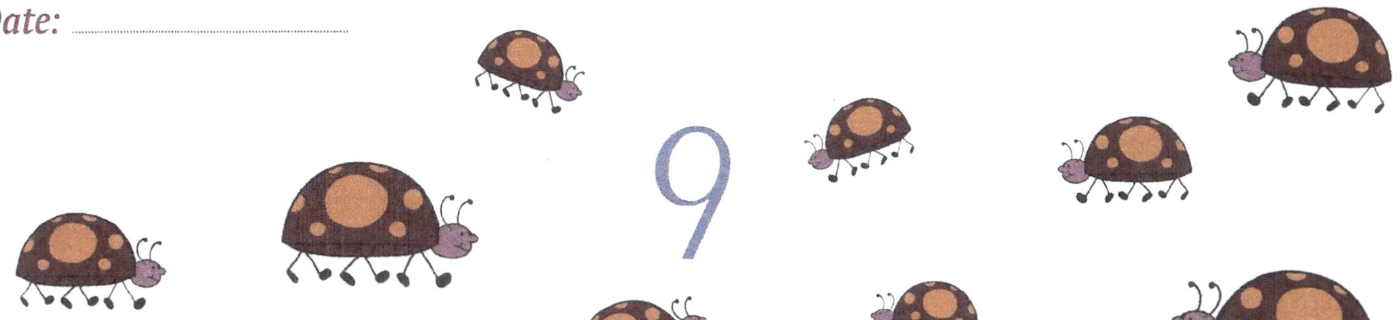

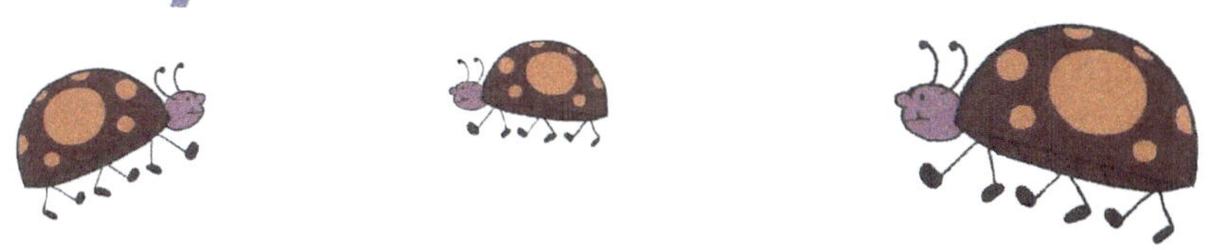

★ Read and write

9

★ Count and colour

7

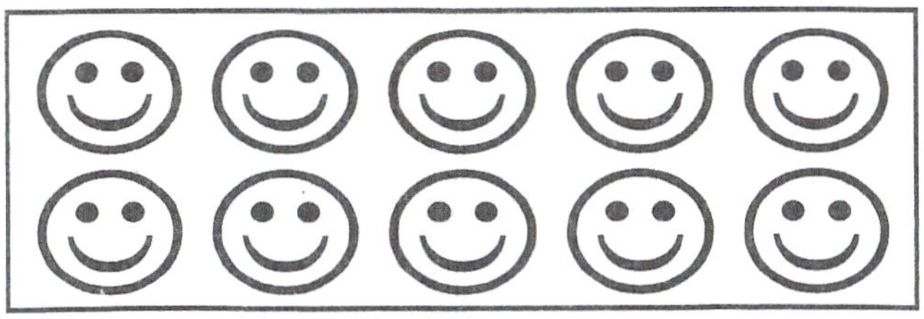

8

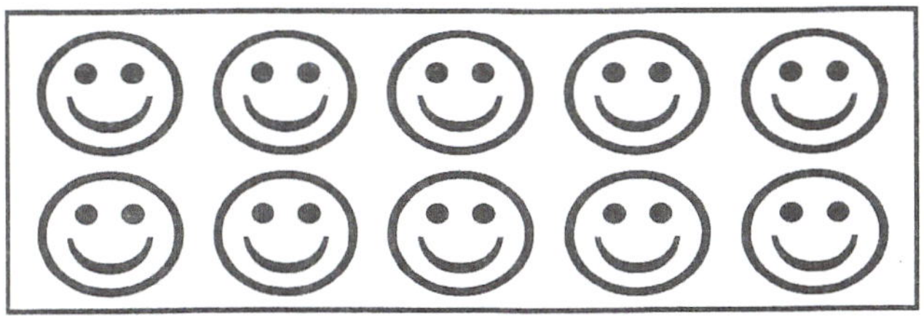

9

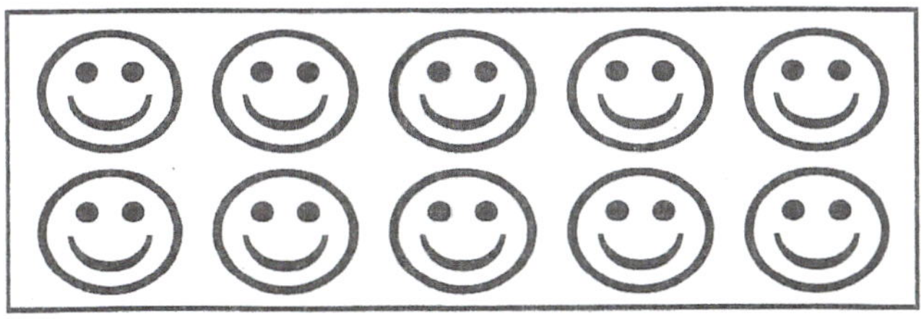

10

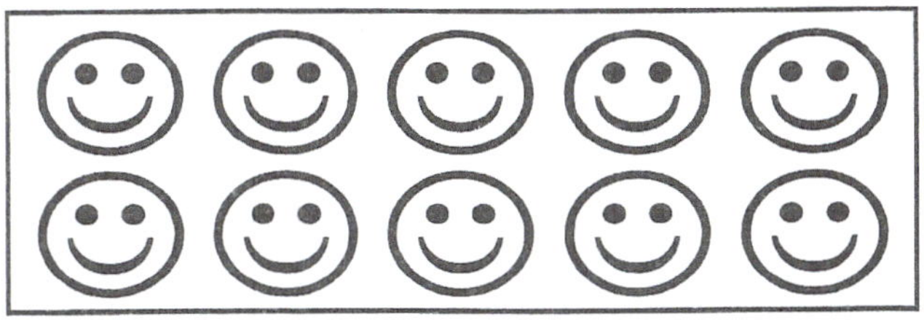

★ Complete

10

★ **Read and write**

10

★ **Match**

9	★★★★★★★★★
10	★★★★★★★★★★
8	★★★★★★
7	★★★★★★★

★ **Count and write**

★ Connect and colour

1•
9•
                                                            •5

                              3
                              •

                        2•   •4

                        8| |6
                        •   •

                              •
                              7

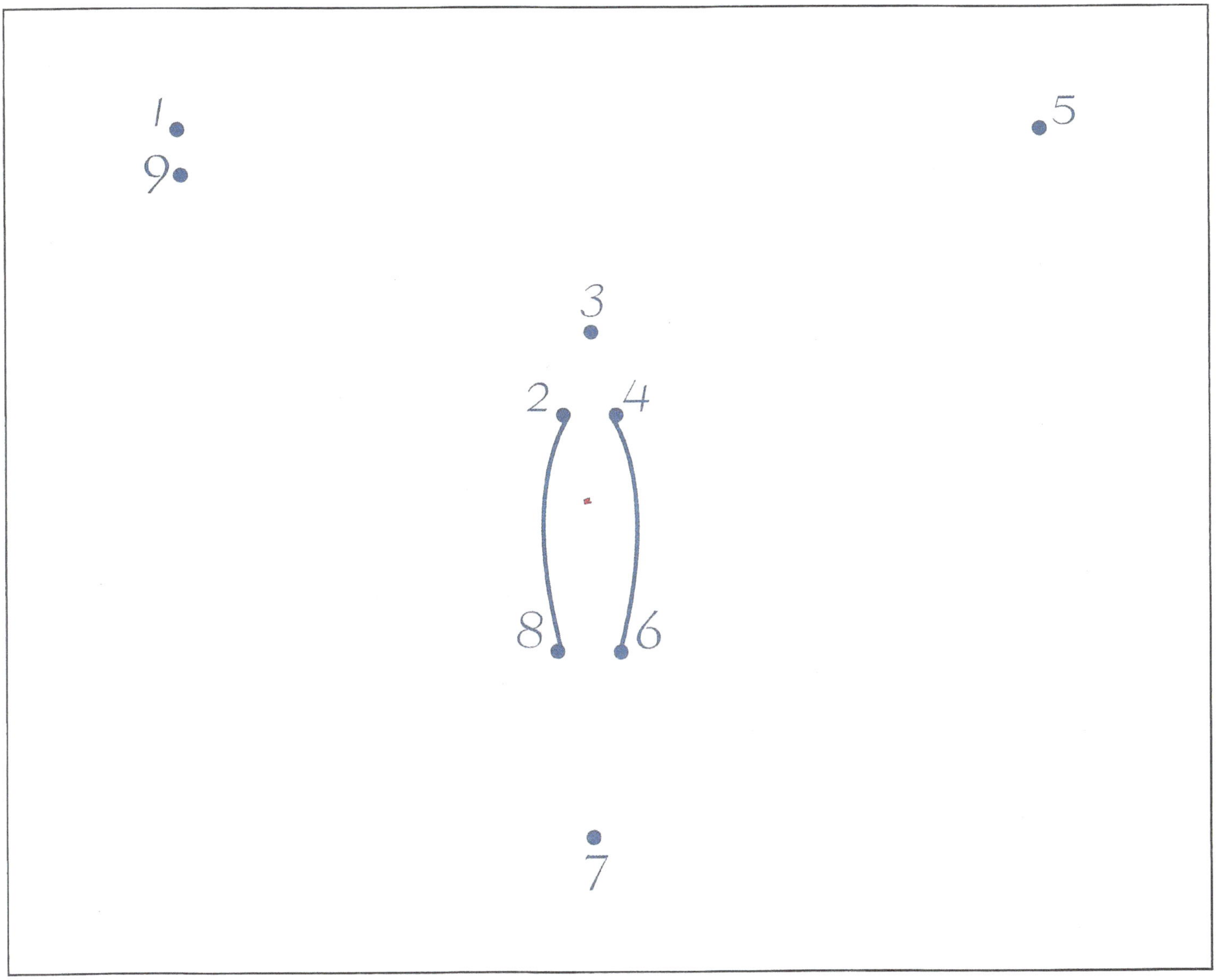

★ Write the missing number

■ 1 [2] 3 [ ] 5

■ 6 [ ] 8 [ ] 10

■ 2 [ ] 4 [ ] 6

■ 5 [ ] 7 [ ] 9

★ Put the numbers
   in the correct order

■ [9] [7] [10] [8]
  [7] [ ] [ ] [ ]

■ [4] [6] [5] [3]
  [ ] [ ] [ ] [ ]

Date: _____

Circle the group with more animals

Circle the greater number

★	4	(6)

★	9	7

★	8	10

★	1	2

★	3	5

★	5	4

★	9	8

★	6	8

★	7	2

★	5	6

★	10	9

★	4	3

★	8	7

★	7	9

★	7	5

★	2	3

Circle the group with fewer flowers

Date: ....

Circle the smaller number

★ ⟨5⟩ 6	★ 9 7
★ 8 10	★ 1 2
★ 3 5	★ 5 4
★ 9 8	★ 6 8
★ 7 2	★ 5 6
★ 10 9	★ 4 3
★ 8 7	★ 7 9
★ 7 5	★ 2 3

Date: ....

Add and write

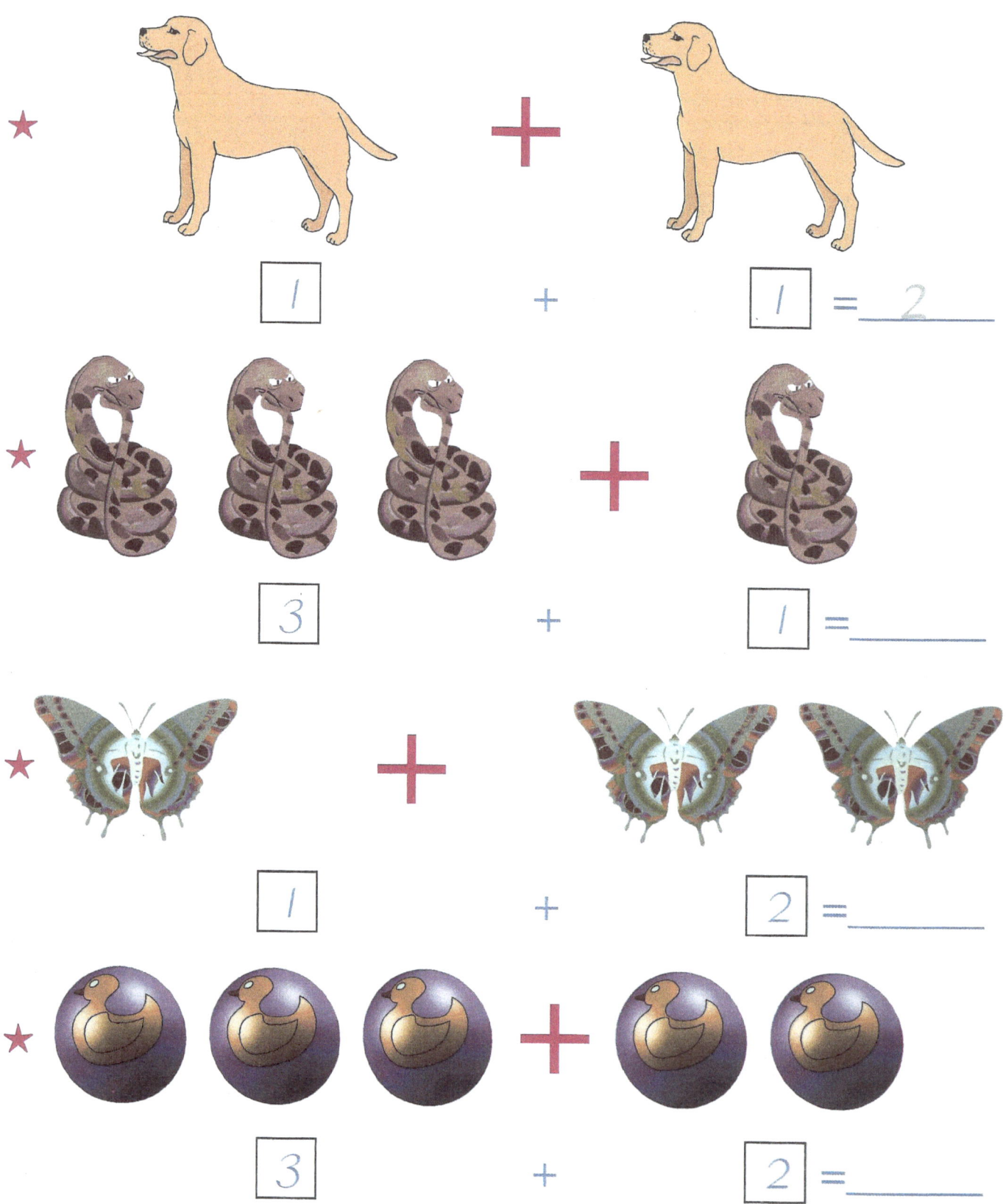

★ ☐ 1   + ☐ 1 = ___2___

★ ☐ 3   + ☐ 1 = _____

★ ☐ 1   + ☐ 2 = _____

★ ☐ 3   + ☐ 2 = _____

Date: ....

★ Read and write

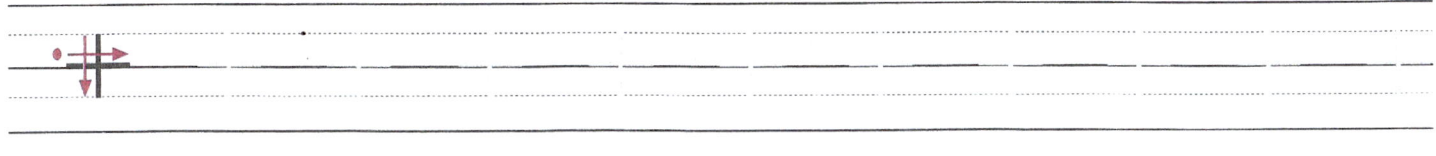

★ Add and circle the correct number

 + 🌹🌹🌹 = 6   8   ⑦

 + 🌼🌼 = 4   3   2

�green�green�green + �green�green�green = 5   7   6

🌺🌺🌺 ☐ 🌺 = 5   8   4

🌷🌷🌷🌷 ☐ 🌷🌷🌷 = 7   8   6

 ☐ 🌹🌹🌹🌹 = 10   8   9

 ☐ 🌹🌹 = 5   4   3

🌹🌹🌹🌹🌹🌹 ☐ 🌹🌹🌹🌹 = 10   7   8

=

★ Read and write

★ Add and write

Date: ....

Add, write and match

★ $3 + 5 = \underline{8}$

★ $1 + 4 = \underline{}$

★ $7 + 2 = \underline{}$

★ $3 + 4 = \underline{}$

★ $2 + 2 = \underline{}$

★ $3 + 3 = \underline{}$

★ $2 + 1 = \underline{}$

Date: ....

Add and write the equation

★  +

2 + 2 = 4

★  +

★  +

★  +

★ +

★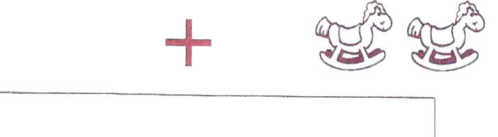

★  +

28

Evaluation

Page Number	Now I can ...	☺	😐	☹	Date	Signature
Page 1	count from 1 to 5					
Page 2	write 1 match pairs					
Page 3	write 2 match pairs					
Page 4	write 3 distinguish between numbers					
Page 5	count from 1 to 5 use a colour key					
Page 6	write 4 distinguish between numbers					
Page 7	count from 1 to 5 create a key					
Page 8	write 5 match pairs					
Page 9	scan a picture create a key					
Page 10	match pairs distinguish between numbers					
Page 11	count from 1 to 10					
Page 12	count from 1 to 8 write from 1 to 6					
Page 13	count from 1 to 6 scan a picture create a key					
Page 14	write from 1 to 7 match pairs					

Page Number	Now I can ...	🙂	😐	☹️	Date	Signature
Page 15	count from 1 to 9 write from 1 to 8					
Page 16	count from 1 to 8 match pairs					
Page 17	count from 1 to 10 write from 1 to 9					
Page 18	write 10 match pairs count from 1 to 10					
Page 19	count from 1 to 9 do a number gap fill put numbers in order					
Page 20	count to 10 differentiate between groups					
Page 21	recognise the greater number					
Page 22	differentiate between groups					
Page 23	recognise the smaller number					
Page 24	add					
Page 25	write the plus sign add					
Page 26	write the equal sign add					
Page 27	add match pairs					
Page 28	write an addition equation